Bald Eagle

Julie Murray

Abdo Kids Junior
is an Imprint of Abdo Kids
abdobooks.com

Abdo
US SYMBOLS
Kids

abdobooks.com

Published by Abdo Kids, a division of ABDO, P.O. Box 398166, Minneapolis, Minnesota 55439.

Abdo Kids Junior™ is a trademark and logo of Abdo Kids.

Printed in the United States of America, North Mankato, Minnesota.

052019

092019

Photo Credits: Alamy, iStock, Shutterstock

Production Contributors: Teddy Borth, Jennie Forsberg, Grace Hansen

Design Contributors: Christina Doffing, Candice Keimig, Dorothy Toth

Library of Congress Control Number: 2018963326

Publisher's Cataloging-in-Publication Data

Names: Murray, Julie, author.

Title: Bald eagle / by Julie Murray.

Description: Minneapolis, Minnesota : Abdo Kids, 2020 | Series: US symbols | Includes online resources and index.

Identifiers: ISBN 9781532185359 (lib. bdg.) | ISBN 9781532186332 (ebook) | ISBN 9781532186820 (Read-to-me ebook)

Subjects: LCSH: Animals--Symbolic aspects--Juvenile literature. | American bald eagle--Juvenile literature. | United States--Seal--Juvenile literature. | Emblems, National--United States--Juvenile literature.

Classification: DDC 929.9--dc23

Table of Contents

Bald Eagle

Look in the sky.

It's a bald eagle!

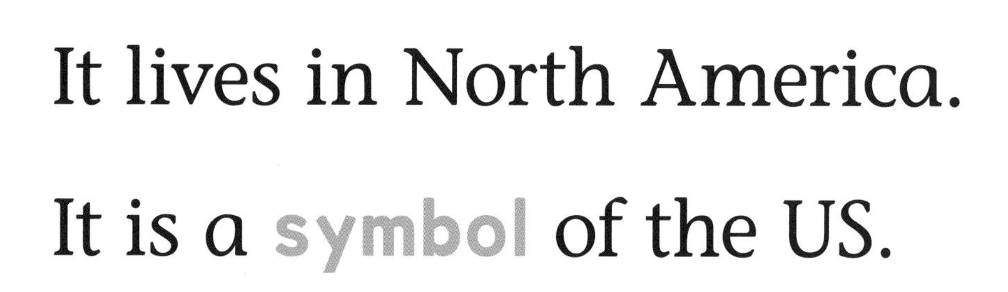

It lives in North America.

It is a **symbol** of the US.

SEAL OF THE PRESIDENT OF THE UNITED STATES
E PLURIBUS UNUM

It is the national bird.

1819

It is a raptor. It stands for strength.

It also stands for freedom.

It soars in the sky.

It is on the US **seal**. This is called the Great Seal.

EMBASSY
E PLURIBUS
UNUM
UNITED STATES OF AMERICA

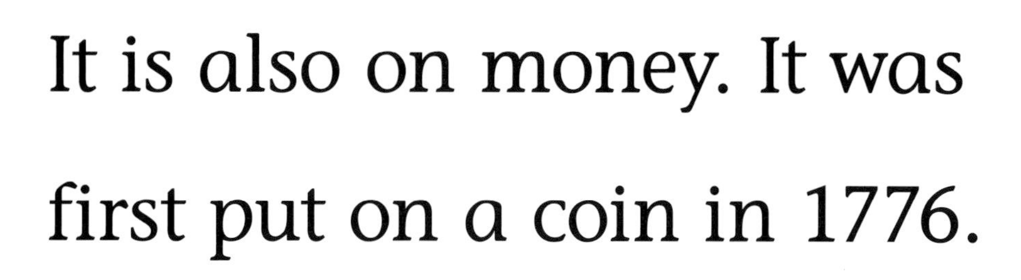

It is also on money. It was first put on a coin in 1776.

THIS NOTE IS LEGAL TENDER
ALL DEBTS, PUBLIC AND PRIVATE
20
FEDERAL
MB 00622070 E
B2
UNITED STATES
FEDERAL RESERVE SYSTEM
THE UNITED STATES OF AMERICA
ONE
ONE
ONE
DOLLAR
UNITED STATES OF AMERICA
E PLURIBUS UNUM
QUARTER DOLLAR
IN GOD
252
112
COEPTIS

It is on stamps too.

United States
Post Office
UNITED STATES POST OFFICE
Main Post Office
Stillwater, MN
Main Post Office
Stillwater, MN
Mon. - Fri. 8:30am - 5:00pm
Saturday 10:00am - 12:00pm
Sunday & Holidays Closed
LIMITED EDITION
STAMPS

Have you ever seen a bald eagle?

United States Seal

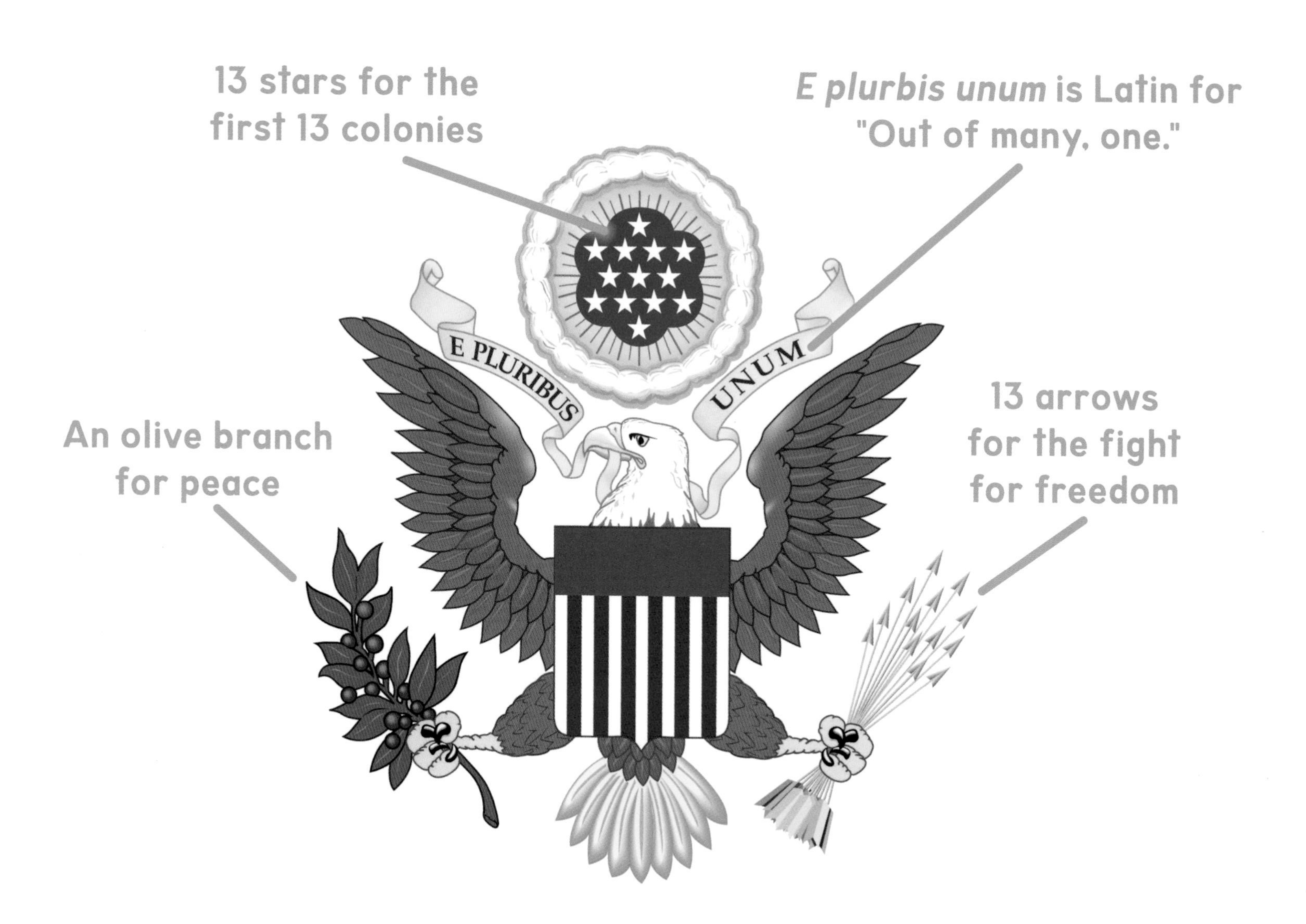

Glossary

raptor
a bird of prey

seal
a design that represents a person or group. Seals are used to make things authentic or official.

strength
the amount of power and influence that a person or group has.

symbol
an object that represents something else.

Index

Visit **abdokids.com** to access crafts, games, videos, and more!

Use Abdo Kids code

UBK5359

or scan this QR code!